How To Plan for Generalization

SECOND EDITION

Donald M. Baer

How To Manage Behavior Series

R. Vance Hall
and
Marilyn L. Hall
Series Editors

pro·ed
An International Publisher

8700 Shoal Creek Boulevard
Austin, Texas 78757-6897

© 1999, 1982 by PRO-ED, Inc.
8700 Shoal Creek Boulevard
Austin, Texas 78757-6897
800/897-3202 Fax 800/397-7633
www.proedinc.com

Library of Congress Cataloging-in-Publication Data

Baer, Donald Merle
    How to plan for generalization / Donald M. Baer.
      p. cm.—(How to manage behavior series)
    Includes bibliographical references.
    ISBN 0-89079-793-5 (pbk. : alk. paper)
    1. Behavior modification. I. Title. II. Series.
BF637.B4B33   1998
153.8'5—dc21   98-38350
                                    CIP

This book is designed in Palatino and Frutiger.

Printed in the United States of America

4   5   6   7   8   9   10   09   08   07   06   05

Contents

Acknowledgments

I owe much of my ability to write this book to my longstanding collaboration with Dr. Trevor F. Stokes. I am also indebted, only a little less intensely, to all my colleagues in the Department of Human Development and Family Life of the University of Kansas, who have striven to share their wisdom with me in all matters of science and behavior; trying to match to their sample has led me into an excellent schedule of reinforcement. A special example is my debt to Mr. Robert Orgel, whose ideas about fluency and generalization contributed greatly to that discussion in the section Aim for a Natural Community of Reinforcement.

Preface to Series

The first edition of the *How To Manage Behavior Series* was launched some 15 years ago in response to a perceived need for teaching aids that could be used by therapists and trainers. The widespread demand for the series has demonstrated the need by therapists and trainers for nontechnical materials for training and treatment aids for parents, teachers, and students. Publication of this revised series includes many updated titles of the original series. In addition, several new titles have been added, largely in response to therapists and trainers who have used the series. A few titles of the original series that proved to be in less demand have been replaced. We hope the new titles will increase the usefulness of the series.

The editors are indebted to Steven Mathews, Vice President of PRO-ED, who was instrumental in the production of the revised series, as was Robert K. Hoyt, Jr. of H & H Enterprises in producing the original version.

These books are designed to teach practitioners, including parents, specific behavioral procedures to use in managing the behaviors of children, students, and other persons whose behavior may be creating disruption or interference at home, at school, or on the job. The books are nontechnical, step-by-step instructional manuals that define the procedure, provide numerous examples, and allow the reader to make oral or written responses.

The exercises in these books are designed to be used under the direction of someone (usually a professional) with a background in the behavioral principles and procedures on which the techniques are based.

The booklets in the series are similar in format but are flexible enough to be adapted to a number of different teaching situations and training environments.

This particular book, *How To Plan for Generalization*, by Donald M. Baer, however, differs somewhat from the other titles in the series. As the author notes in the list of concerns, which he presents early in his text, this manual is

---

R. Vance Hall, PhD, is Senior Scientist Emeritus of The Bureau of Child Research and Professor Emeritus of Human Development and Family Life and Special Education at the University of Kansas. He was a pioneer in carrying out behavioral research in classrooms and in homes. Marilyn L. Hall, EdD, taught and carried out research in regular and special public school classrooms. While at the University of Kansas, she developed programs for training parents to use systematic behavior change procedures and was a successful behavior therapist specializing in child management and marriage relationships.

really not for the beginning practitioner who has no experience in systematically bringing about behavior change. Most of the other books in the series more or less can stand alone. Though they are meant to be used under the direction of a professional, they present a specific procedure or set of procedures with a set of exercises in a simple, step-by-step fashion so that upon completion the practitioner should be able to understand and use that procedure effectively.

To make use of *How To Plan for Generalization,* however, one needs to be already somewhat skilled and knowledgeable about the procedures presented in the other books of the series to understand and benefit from this text. Furthermore, it is somewhat different in format. Although it presents specific steps and frequent, clear examples of the points it makes about generalization, the reader is not led to carry out the procedures with a specific clientele. One reason for this is that the subject of the book is complex, and it would take a much larger book or a series of booklets to achieve that goal. More importantly, the reader is led to see how much of what kind of generalization is wanted in each specific case, because this varies from case to case.

What this book does is to introduce the reader to the topic of generalization. It provides a comprehensive look at the major dimensions of generalization and alerts the reader to the need to be aware of those dimensions if one expects behaviors to be usefully displayed and in contact with natural reinforcement communities. It is essential that persons instructing practitioners using other books in this series have a good understanding of its content, for generalization issues will inevitably arise as people attempt to bring about systematic behavior change.

We would also advise that those who have become skilled in the procedures presented in the other booklets in the series study it. Perhaps it should be read with *How To Maintain Behavior* as a final chapter of the series so that they will be aware of and able to plan for these important dimensions of behavior change.

Another aspect of this book that differs somewhat from the others in the series is that the revision is essentially unchanged from the original version. This is so in part because the field itself has changed little since it was first written 10 years ago. It is also due to the fact that the author, considered the leading authority and the founding father of this area, presented such a precise, incisive original survey that the editors could find little that needed updating or change. That is also why this remains one of the most definitive statements about generalization available in the literature today.

As always, we invite your comments, suggestions, and questions. We are always happy to hear of your successes in changing your own behaviors and the behaviors of other persons to make your lives more pleasant, productive, and purposeful.

R. Vance Hall &
Marilyn L. Hall
Series Editors

How To Manage Behavior Series

How To Maintain Behavior

How To Motivate Others Through Feedback

How To Negotiate a Behavioral Contract

How To Plan for Generalization

How To Select Reinforcers

How To Teach Social Skills

How To Teach Through Modeling and Imitation

How To Use Group Contingencies

How To Use Planned Ignoring

How to Use Positive Practice, Self-Correction, and Overcorrection

How To Use Prompts To Initiate Behavior

How To Use Response Cost

How To Use Systematic Attention and Approval

How To Use Time-Out

How To Use Token Economy and Point Systems

Introduction

This book is written for persons who have the task of bringing about deliberate behavior changes in themselves or others. Anyone may want to accomplish a personal behavior change—to weigh less, to work more, to smoke less, or to socialize more. Many persons are charged with making behavior changes—parents in their children, teachers in their students, counselors in their clients, institutional staff in their residents, and employers in their employees. And many persons choose to support behavior changes in each other—marital partners, lovers, coworkers, colleagues, and friends.

Persons who know how to bring about deliberate behavior change in themselves, or in others, also know that, at first, the change may not be widespread enough to be truly valuable. A child with poor articulation is taught better speech by a therapist in a clinic and speaks better to the therapist there, but continues to show poor articulation elsewhere with other persons. An overweight man reduces his eating to within desired limits in his home, but he still overeats in restaurants, at parties, in friends' homes, and at movies and sporting events. A mother easily teaches her son to say "please" when asking for food at the table, and "thank you" when receiving a gift, but she is dismayed when he does not say "please" when asking for help or "thank you" when help is given.

In these cases, the persons causing the behavior change needed a more widespread effect than they have created directly. They probably expected that the changes they had made directly would be enough, so that a "concept" or "rule" had been taught, a "skill" learned, or a "habit" formed—and that the other desired behavior changes would follow naturally. In these cases, they were disappointed. They probably wondered if they had taught poorly, and, more than likely, they suspected that their student was not very bright. (The man trying to eat less probably concluded that he had no "will power.")

This book offers all such persons two lessons—a passive one and an active one.

The passive lesson is that teaching one example never automatically instills a concept, a rule, or a habit. Learning one aspect of anything never means that you will automatically know the rest of it. Doing something skillfully now never means that you will always do it well. Resisting one temptation consistently never means that you now have character, strength, and discipline. Thus, it is not the learner who is dull, learning disabled, or immature, because all learners are alike in this regard: *no one learns a generalized lesson unless a generalized lesson is taught.*

Donald M. Baer is a Roy A. Roberts Distinguished Professor of Human Development and Family Life. He is internationally known for his extensive research and theoretical contributions to the field of applied behavior analysis. The senior editor of this series is also proud to acknowledge him as his mentor and friend.

The active lesson is that there are ways to teach a generalized lesson, or at least encourage one to be learned. To approach the active lesson, we need to always think of what we teach in terms of these sequential questions:

1. What lesson or lessons are we going to teach directly?

2. Does what we want the student to learn stop with what we are going to teach directly, or does it need to be more widespread and generalized than the examples we will teach?

3. If what we want the student to learn needs to be more generalized than the examples we intend to teach, are prepared to teach, know how to teach, are willing to teach, or have time and materials enough to teach, then what are all the ways in which the lesson needs to be more generalized?

4. Considering all the ways in which we want this lesson to be generalized, what are the most likely methods for teaching so that the result is a sufficiently generalized learning?

The answers to question 4 are the major business of this book. But do not lose sight of the fact that before you can effectively use the answers to question 4, you need to answer question 3 as carefully, accurately, and thoroughly as possible. An answer to question 3, in turn, first requires answers to questions 1 and 2. The answer to question 3 usually will determine the answers to question 4; and failure to answer question 3 usually will mean that you are only throwing dice when you try to choose a method for generalizing what you are going to teach.

This discussion is introductory; it needs examples to become useful. But before those examples are considered, some background is in order.

Some Concerns

Concern 1

This book is written for persons who already know how to make systematic, deliberate behavior changes. If that is not true of you, then you are reading the wrong book. Trying for generalized behavior changes is something you need to learn very soon *after* you learn how to make behavior changes, or, if you learn concurrent lessons well, *as* you learn how to make behavior changes. But it is useless to try to learn these methods *before* you have learned how to make behavior changes. Sometimes, it may be worse than useless. As road signs in some very difficult parts of the British Commonwealth occasionally state, "You Have Been Warned!"

How do you know if you already know how to make some kinds of systematic and deliberate behavior changes? The fact that you have sometimes succeeded in teaching something to someone is definitely not evidence that you know systematically how to do it. The crucial test is whether or not you always know how to accomplish change, and whether or not you know just how you will go about future teaching so that success is assured.

Systematically is the key word in this concern. Persons who understand and practice behavior change systematically know the principles that describe and explain what they do; understand the techniques embodying those principles; are able to invent new variations of those techniques when necessary; are skilled in ways to observe, measure, record, and graph not only their own behaviors as they teach, but also their students' behaviors as they learn what is taught; and are thoroughly familiar with the designs that will show whether or not it was their teaching that accomplished whatever behavior changes occurred. If any of this sounds at all mysterious, then your reading of this book is premature and should stop here. The correct alternative is university-level instruction in behavior analysis and applied behavior analysis, or at least total knowledge of, competence in, familiarity with, and consistently successful practice with the two-part Managing Behavior series (Axelrod & Hall, 1999; Hall & Van Houten, 1983) and the entire How To Manage Behavior series that follows them. (Titles in the How To Manage Behavior series are listed in this text's Preface to Series.)

This book and its procedures do not stand by themselves. That is the first and most important concern, and it cannot be overemphasized.

Concern 2

This book might have been titled, *How To Try for Generalization. Try* is the word that expresses this concern. This book will describe some procedures that can accomplish more widespread behavior changes than are caused directly. That does not mean that the procedures will always succeed. It means only that they can. The manifest truth, embodied in the experience of those who report research in the field of behavior analysis, is that these techniques sometimes fail to create a more generalized effect than was already in hand before they were applied, or than would have been the case even if they had not been applied. These are essentially recipes. Like most recipes, they can fail. When recipes fail, it is usually because some condition essential to their success is not stated explicitly as part of the recipe, and some cook, not knowing that this condition was essential and not finding it listed in the recipe itself, failed to include it in the recipe's realization. And so the dish failed to be edible, usable, or attractive.

The situation here is even more difficult. It is *not* that applied behavior scientists know all the conditions and ingredients that make a generalization recipe work, but sometimes forget to write them down. It is *not* that they sometimes assume that their readers also know and therefore do not need to be told. The problem is that none of us knows all the ingredients and conditions necessary and sufficient to the success of these recipes. The information that we can pass on is a description of procedures and conditions under which these recipes often have worked, and, in some cases, almost always have worked. That is the best that we can do. Thus, the best that you can do with these techniques is try for generalized behavior changes and, if you fail, to reconsider the problem and perhaps try again. In reconsidering the problem, you will be troubleshooting, and you will need (as every troubleshooter needs) all the knowledge and lore available if you are to do it with the best probability of success. Remember Concern 1?

Concern 3

These recipes are not analyses of generalization. In fact, what they accomplish (when they work) can be called generalization only for practical purposes. At the level of scientific analysis, other terms (sometimes "stimulus control," sometimes "natural communities of reinforcement," sometimes "intraverbal mediation") or combinations of terms more often would be correct. A change in terms usually means a change in process, and that is almost certainly the case here. These are not procedures that serve a single process called "generalization," and that get it to happen in the real world. These are procedures that imperfectly nudge some variety of processes into action. Since they do not in themselves make clear what process or processes they are evoking, it will be difficult to adjust them to make the relevant process or processes work better.

Then why do they work at all? Apparently, they do prod some relevant behavior–analytic processes into action, even if imperfectly, even if inefficiently, and even if ignorantly. These procedures are not collected here to improve our understanding of generalization, stimulus control, intraverbal mediation, or the ecology of natural communities of reinforcement. They are collected here because many of us need to produce more widespread behavior changes than we can make directly, and these procedures often work (for whatever reasons) to do just that. We can use them without understanding why they work. And because they serve a single, pragmatic goal for us, we can think of them as a single class of techniques—generalization techniques. Pragmatically, they are generalization techniques; fundamentally, they are not. But this book is not about fundamentals; it is about pragmatics. The essence of this concern is that we be pragmatic knowingly, and that we realize that a great deal of analytic research relevant to our single, pragmatic goal has not yet been done.

Concern 4

This book is not only a book. It is also an invitation to accomplish some applied behavior analysis with a set of techniques already acknowledged as not very analytical. Most importantly, it is an invitation to apply those techniques to real persons needful of some behavior changes of a more general character than what we think we can accomplish directly. Applied behavior analysis usually is taught at the university level, for many hours, with many thousands of pages of scientific and research-based experience to read, and with a good deal of practice supervised by researcher–practitioners who themselves are thoroughly experienced. Almost none of that is going to happen. Instead, a very small number of pages will be read, and, as a result, some generalized behavior changes may be attempted. The essence of this last concern is that reading a small recipe book may not be an effective enough method to teach anyone to do well. Yet, it might be. Whether it is or whether it is not is an empirical question that requires research to be answered in any general way. The research can hardly be done without the book. Given the book, the research then can be done—but only if the book's users use it carefully, systematically, with good measurement, and in useful research designs, and if they report their results. If the users of this book meet the specifications of Concern 1 above—if they are expert in both parts of Managing Behavior and the How To Manage Behavior series—then they will indeed do and report the necessary research. Thus, Concern 4 is again Concern 1. Please reread Concern 1 now. And, as your experience grows, please write to me about it (with graphs, of course).

Some Active Lesson Examples and Procedures

The active lesson of this book is that there are ways to make behavior changes that encourage a more generalized change than we can make directly. Some of those ways will be described here. But the major point now is that before those ways are presented, you need to learn how to approach them. A good approach lets you choose a way, or a set of ways, with the best probability of being effective. The approach has been sketched already; it is the four sequential questions of the introduction. Now they need elaboration, and the elaboration has to be yours. You are the person who knows what behavior change you intend to accomplish, and how generalized the changed behavior needs to be. This book cannot know that. It can only invite you to write a plan that will make the remainder of this book useful.

Your plan must start with the answer to the most basic question:

Exactly what behaviors are you going to change directly, and in whom?

In other words, *who is the student, what exactly is the lesson,* and *who is going to teach it?* Sometimes, you are the teacher, as well as your own student. (Remember the man who wanted to eat less?) Sometimes, you are the teacher, and someone else is your student. (Remember the mother teaching her son to say "please" and "thank you"?) Sometimes, you are the teacher's consultant. (Suppose that the mother asked you how to get her son's "please" and "thank you" behavior generalized?)

(pencil icon) _____

Think through your case, and list all the persons whose behavior must change: student, teacher, consultant. Certainly you are one of those persons.

_____

_____

_____

▶ **Step 1: The Student**

Begin with the student—the target of the behavior change. First, list all the behaviors that need to be changed in the student. This will not be an easy list to write, but the effort that it will require is worthwhile. When you have finished, you will have the beginning of a much more realistic plan than would be possible any other way—and a much better prospect of success.

Think through *all* the behaviors that need to be changed in the student.

(pencil icon) _____

List all the forms of the student's behavior that need to be changed.

_____

_____

_____

_____

Example: The target behavior is to say "please" and "thank you." Is "please" enough, or should you also teach "Would you please . . ."? Should you teach "please" or "Would you please" as the first word(s) of every request, or as the last word(s) of every request? Or some of both? Is "thank you" enough, or should you also teach "Thanks," "Mmmm,"

(continues)

"Wow!," "Thank you so much," "Just what I wanted," and so forth? Should a smile go with all of these forms? With some of these forms?

_____

_____

_____

Example: The target behavior is a misarticulated "l" sound. Usually, the speaker substitutes a "w" sound for it. The proper articulation of "l" depends somewhat on whether it is the initial sound, the final sound, or a medial sound in a word; on whether it is preceded by a vowel or by a consonant; and whether it is followed by a vowel or by a consonant.

Should you teach all of these forms?   ☐ Yes ☐ No   Why?

_____

_____

_____

_____

Example: The target behavior is dressing oneself. The forms of this behavior can include putting feet through holes, feet into and through tubes (socks and pants legs), feet into shoes, and thongs between toes; hands through holes and hands into and through tubes (sleeves); head and arms into and through the bottoms of garments, and feet and legs into and through the tops of garments; right arm, right leg, and right foot into right sleeves, legs, and shoes; left arm, leg, and foot into left sleeves, legs, and shoes; buttoning; tying; lacing; snapping; hooking; binding; wrapping; zipping; and squeezing into and out of.

Should you teach all of these forms?   ☐ Yes ☐ No   Why?

_____

_____

_____

_____

_____

(continues)

List all the situations, settings, and places where the needed behavior changes in the student should occur, as well as all the persons with whom it should occur.

_____

_____

_____

> **Example:** The target behavior class is "please" (in all the forms that you decided were necessary). List all the kinds of requests that this child ought to accompany with some form of "please" when asking for servings at the table, when asking for help, and when asking for information. Should every request that the child ever makes have a form of "please" attached to it? If not, then can you define the kinds of requests that should have "please" attached to them, and in such a way that the child can distinguish them from the kinds that do not need a "please"? In fact, can you define the concept of "request" so that you can teach it to a child? This is important. If you cannot define it, you probably cannot teach it effectively, and you are likely to be disappointed with what the child learns from your attempts. (This is not to say that each of us has not learned a complex set of situations that do and do not call for some form of "please," or that we are not satisfied with the way that these behaviors work in us. It is to say that if we cannot specify how those behaviors work in us, we will have trouble teaching them to our child so that the behaviors work in just the same way—short of correcting every instance of the behaviors in the child that sounds wrong to us.)

_____

_____

_____

> **Example:** The target behavior class is dressing oneself. Whatever forms of this behavior class are taught, it will usually be best if learners dress themselves in only their own clothing, and not garments of their siblings (in a family) or their roommates (in a cooperative living arrangement) or their wardmates (in an institution). It is also important that they dress themselves in any new garments. List the situations that define when self-dressers should and should not put on the clothes that are present. Do not forget that dress should match the weather, the nature of the next activities, and cultural norms.

(continues)

_____

_____

_____

Example: The target behavior is a correctly articulated "l" sound. List the situations, settings, and persons where the correctly articulated "l" should be spoken. Are there any where it should not? (This is one of those examples in which it seems desirable that the behavior change should be displayed everywhere and all the time—except when playing a character with a speech defect in some drama, perhaps.)

_____

_____

_____

Example: The target behavior is for institutionalized children with mental disabilities to greet the ward staff (because consistently greeting the staff will make the children seem more human to the staff, and thus more deserving of social and teaching inputs of all kinds). The form of the greeting is restricted to a wave and a smile, in view of the children's very limited abilities.

List the persons who should be greeted, and those (if any) who should not (such as total strangers). List the situations in which greetable persons should be greeted, and those in which they should not. (After an absence, of course, but an absence of how long? One hour? Sure! One minute?)

_____

_____

_____

_____

_____

Example: The target behavior is eating less. The situations in which eating should be reduced are meals, parties, friends' homes, movies, sporting

(*continues*)

events, picnics, and friendly offers to share by others who are eating. List all other situations in which food is available and should be resisted wholly or partially. And consider another situation that almost certainly will occur in any such diet program—occasional loss of self-control, resulting in an eating binge. The binge is a happening that many dieters react to as if their entire program had now failed and so was canceled. Thus, the binge needs to be included as a "situation" that will happen, and it is one after which reduced eating behavior is again the immediate target behavior.

_____

_____

_____

_____

_____

_____

_____

Possible Benefits

There are six possible benefits of listing all the forms of the behavior that are to be changed, and all the situations in which it is to be changed or not changed:

1. You now see the full scope of the problem ahead of you and, thus, see the corresponding scope that your teaching program needs to have.

2. If you teach less than the full scope of the problem, you will do so by choice rather than by forgetting that some other forms of the behavior existed and could be important, or that there were some other situations in which the behavior change should or should not occur.

3. If a less than complete teaching program results in a less than complete set of behavior changes, you will not be surprised.

4. You can decide to teach less than there is to learn, perhaps because that is all that it is practical or possible for you to do.

5. You can decide what is most important to teach. You can also decide to teach the behavior to them in a way that encourages the indirect development of some of the other forms of the desired behavior, as well as the indirect occurrence of the behavior in some other desired situations in which it should happen, but that you will not or cannot teach directly.

6. But, if you choose the option discussed in number 5 above, rather than the complete program implicit in number 1, you will do so knowing that the desired outcome would have been more certain had you taught every desirable behavior change directly. The best that you can do, otherwise, is to encourage the behavior changes that you do not cause directly. So, you will have chosen the option in number 5 either of necessity or else as a well-considered gamble after a thoughtful consideration of possibilities, costs, and benefits.

▶ Step 2: Other Persons

Now that you have listed all the forms of behavior change in the student that should occur, and all the situations in which the student's behavior changes should occur, do the same thing for everyone else involved.

1. Who is teaching these behavior changes? What behaviors must they display to teach these behavior changes? How many forms must those teaching behaviors take? In what situations must they occur or not occur?

_____

_____

_____

2. If these behavior changes start to occur in all the forms that are desired and in all the situations that are desired, who else will be affected? Who lives with the learner? Who works with the learner? Who plays with the learner? Who is an audience for the learner? Who contacts the learner?

_____

_____

_____

This point is extremely important to any behavior change, and most especially to a behavior change that needs to be more generalized than its direct teaching will make it. All persons are potential teachers of all sorts of behavior changes. Just because you are designated as "teacher" or "behavior analyst" does not mean that you have an exclusive franchise on the ability to make deliberate behavior changes. In fact, there is no possibility of such a franchise. Everyone in contact contributes to everyone else's behavior, both to its changes and to its maintenance. The only aspect of this interaction that is not universal (yet) is the formalized knowledge of how it works. (Although, books like this one, and many others, are aimed at extending that franchise indefinitely.)

At the very least, everyone else affected by or contacted by the behavior changes that you intend to teach must tolerate those changes in the learner. At best, they may be involved in the teaching, the maintenance, and the generalization of these changes.

Thus, you now need to make two more lists:

1. If everyone else affected or contacted by these behavior changes is merely to tolerate the behavior changes, and not act against them, what behaviors must they display or stop displaying? How many forms must their behavior changes take? In what situations must their behavior changes occur?

2. If everyone else affected or contacted by these behavior changes is to actively support the behavior changes, and not merely tolerate them, what behaviors must they display or stop displaying? How many forms must these behavior changes take, and in what situations must these behavior changes occur?

(continues)

Example: A nonambulatory, institutionalized girl with mental disabilities was always moved from place to place in a wheelchair pushed by supportive ward aides. A behavior analyst taught the child to walk on crutches quite successfully, but found that this skill did not generalize in the behavior analyst's absence. The girl did not use the crutches and was still pushed in a wheelchair by the ward aides. The behavior analyst then convinced the ward aides that it was to the child's benefit for them to stop pushing her chair and to point to her crutches at such moments. They agreed and did just that. The child promptly began using crutches to walk everywhere in the institution. Within a few weeks, she had much greater mobility and range of travel than ever before. That mobility was supported and extended by the affectionate, approving reception she received from ward aides when she walked to their current workplace to see them.

More Possible Benefits

There are six possible benefits of listing all the forms of behavior change and all the situations in which these behavior changes should occur for everyone else involved. They are the same six benefits that were gained by listing the forms and situations of behavior changes for the student. Read them again now, noting the additions that this discussion requires:

1. You now really see the full scope of the problem ahead of you, and thus see the corresponding scope that your teaching program needs to have.

2. If you teach less than the full scope of the problem, you will do so by choice rather than by forgetting that some other forms of the behavior existed and could be important; or that there were some other situations in which the behavior change should or should not occur; or that there were some other persons whose behavior changes were also involved, and that their behavior changes had forms and situations to be considered as well.

3. If a less than complete teaching program results in a less than complete set of behavior changes, you will not be surprised.

4. You can decide to teach less than there is to learn, and to fewer people than are involved, perhaps because that is all that it is practical or possible for you to do. But, if you do that, you must remember the next point.

5. You can decide what is most important to teach and who is most important to teach. You can also decide to teach the behavior to them in such a way that encourages the indirect development of

some of the other forms of the desired behavior, as well as the indirect occurrence of the behavior in some other desired situations in which it should happen, but that you will not or cannot teach directly.

6. But, if you choose the option discussed in number 5 above, rather than the complete program implicit in number 1, you will do so knowing that the desired outcome would have been more certain had you taught every desirable behavior change to everyone involved, directly. The best that you can do, otherwise, is to encourage the behavior changes that you do not cause directly. So, you will have chosen the option in number 5 either of necessity, or else as a well-considered gamble after a thoughtful consideration of possibilities, costs, and benefits.

▶ Step 3: Selecting the Teaching Option

Now that you have listed all the behavior changes that should occur, in all their forms, in all their situations, for everyone involved, decide on either number 1 (teach all the needed behavior changes directly) or on number 5 (teach less than is necessary, but teach the most important ones first, and teach in a way that will encourage the development of the untaught behavior changes).

If your choice is the certain, but more difficult and sometimes impossible, option number 1, you need go no further in this book. If your choice is the less certain, but perhaps more realistic, option number 5, then do the two things that option requires:

First, decide what is most important to teach directly and to whom it should be taught.

This book cannot help you with that problem, not knowing what your problem and its situation may be. But if you are still reading this book, you should have met the specifications set out in Concern 1 (in the second section of this book). You already know systematically how to make deliberate behavior changes. Then you should be able to make decisions about what to teach and to whom with some degree of wisdom. More importantly, you should know how to check on the soundness of your decisions and to alter them if they are not working out well.

Second, in light of what you have decided not to teach directly, and therefore what needs to be generalized and in whom, you now need to consider the classes of generalization techniques for incorporation into your teaching.

Some Generalization Procedures and Examples

▶ **Step 4: Aim for a Natural Community of Reinforcement**

The everyday environment is full of steady, dependable, hardworking sources of reinforcement for almost all the behaviors that seem natural to us. That is why they seem natural.

The way that we are formed, the way that gravity operates, and the way that the surface of this planet is arranged all interact to make walking and running always effective, useful, and seemingly easy ways to get from here to there. At any moment, many of our reinforcers are *there* rather than *here*, and so any means of getting to *there* will be reinforced. Walking and running are two of those means that happen to fit us, the world, and the laws of nature if they are done just right. Look at the behavior of walking and running, and you will see that it is the precise use of the legs relative to the rest of the body that succeeds. Less precise, less stereotyped uses of the legs will quite likely have us fall or wander from our reinforceable direction. Who taught that precise, wonderfully skillful performance? Parents often urge their young children into walking, help hold them up, and reinforce locomotion enthusiastically as it develops—but parents do very little to give walking its precision. Most parents do not understand the precision of motor development. The natural community of reinforcement does the real teaching job by punishing mistakes, ignoring irrelevant responses, and reinforcing correct performances by letting the child get from *here* to *there* (where the other reinforcers are). The generalized skills of walking and running are a much larger set of behaviors than any parent can encourage in a child, yet the natural community is quite capable of fulfilling that very large teaching assignment.

The walking and running example is typical of natural reinforcement communities. Natural reinforcement communities not only support already made behavior changes that fit them but they also elaborate, extend, polish, refine, and perfect those changes, as well as maintain them indefinitely. They are the perfect allies for anyone wanting to produce a generalized behavior change.

Thus, most of us, once taught, continue to read, and to read better and better, for the rest of our lives. The content of what we read can be full of reinforcers for us, and the information in what we read can let us manage certain parts of the rest of our lives to increase reinforcement and minimize punishment.

Most of us, once taught to talk, continue to talk to and listen to others, and to do both better and better, for the rest of our lives. The content of what we hear can be full of reinforcers for us, and the ability to control certain aspects of our social world by talking to it increases many sources of reinforcement and decreases many sources of punishment.

This kind of logic has led numerous applied behavior analysts to argue that the only behavior changes that we should make deliberately are those that meet and interact with natural communities of reinforcement. It may require our intervention to begin a behavior change, but if natural communities of reinforcement do not exist to maintain, extend, and refine it, then either we must continue to do all of what we want done, or we must be willing to see that behavior change disappear.

A good rule is not to make any deliberate behavior changes that will not meet natural communities of reinforcement. Breaking this rule commits you to maintain and extend the behavior changes that you want, by yourself, indefinitely. If you break this rule, do so knowingly. Be sure that you are willing and able to do what will be necessary. This is not just a good rule, but perhaps the best rule in this book and in this series.

Avoiding Mistakes

There are two mistakes to avoid when following the natural communities rule.

1. It can be a mistake to think that there is no natural community of reinforcement for a particular behavior change. The problem may be simply that you did not teach the change well enough for it to contact its natural community.

2. It can be a mistake to think that there is no natural community of reinforcement for a particular behavior change. The problem may be simply that the natural community is asleep and needs to be awakened and turned on.

Example: If you teach a child to read, for example, continue the teaching until the child can read with some fluency. If reading is difficult, slow, and very uncertain work, the child cannot contact the natural community of reinforcement in reading. Reinforcement is inherent in the meaning of what the child reads, and, if reading is done poorly, little meaning is gained. In several studies of token reinforcement systems in remedial classrooms, children were given tokens for practicing reading, arithmetic, and spelling. As they gained fluency in each subject, occasional discontinuations of the token reinforcement were tried to see if any academic behaviors would continue without it. At first, all academic behaviors stopped when reinforcement of them stopped. But, as reading gained in fluency, stopping the reinforcement gradually lost effectiveness. Finally, the children read just as much without reinforcement for reading as they did with the reinforcement for reading. Shortly after that, it was found that the children would pay their hard-earned tokens for the privilege of reading. They showed no such willingness to pay for opportunities to do arithmetic or spelling. Thus, when reading was done well enough, it met a natural community of reinforcement;

the other topics did not. Or, at least, performance of the other topics had not yet become fluent enough to meet a natural community of reinforcement, if one existed for them. There are persons who act as if the opportunity to do mathematics problems meets some natural community of reinforcement. Uniformly, they are very, very good at mathematics. We often suppose that the reason that they are so good is that they find mathematics reinforcing. Perhaps the true reason that they find it reinforcing is that they are so good at it! In more technical terms, perhaps fluency preceded generalized maintenance.

Behavior changes that seem to need generalization may only need better teaching. Try making the students fluent, and see if they still need further support for generalization. Fluency may consist of any or all of the following: high rate of performance, high accuracy of performance, fast latency given the opportunity to perform, and strong response.

Example: Children who were not doing very good work in their classrooms were taught to raise their hands, get their teachers' attention, and then ask the teachers to look at the work that they had just finished. They were taught to do this only when they thought that they had done good work. The teachers' natural response was to admire and approve of correct work and to point out errors in incorrect work and ask that it be done again. Children who consistently called their teachers' attention to their good work improved substantially in the rate and accuracy of that work, and they doubled their teachers' rates of praising them. When the children did not recruit their teachers' praise for their good work in this way, the work was poorer and the teachers' likelihood of attending to good work was much lower. The teachers should have been natural communities of reinforcement for good work by their students, but without help they hardly deserved that label. In a way, the teachers were a dormant natural community, asleep at the switch. Awakened, they were very good. It was the children who awakened them, simply by a low but consistent rate of recruiting the teachers' attention to their work when it was probably correct.

We have very little experience with examples like these, yet they are very encouraging. We ought to try them more often. Two points are worth special caution:

1. The children were reluctant to use their recruiting skills on their teachers even after they had been taught them well enough to be fluent in them. Since they did not use them naturally, they had to be reinforced (by an outside-the-classroom person) every day for using these techniques. When that was done (quite easily), they did use recruiting skills consistently and effectively, as just described. You would think that these recruiting techniques would meet a natural

community of reinforcement themselves—the same community that they had just awakened for academic work. Perhaps in time they do. Research simply has not gone far enough yet to check that possibility. Meanwhile, be prepared to support newly taught natural community recruiting skills for awhile.

2. Too high a rate of attempting to recruit natural communities of reinforcement may not recruit them; instead, it may irritate other persons and awaken them as natural communities of punishment. At the other extreme, too low a rate of recruiting natural communities may awaken them so little that they hardly function. Just the right range of rates may have to be found, and what that range is no doubt will vary from problem to problem. In these applications, typical teachers were asked first to estimate a rate that would just begin to bother them, and care was taken to keep the children's recruitment rates well below that ceiling. Something just like that probably should be done in every application, until we have enough experience to see what generalizations (of our own) are possible. Think through and check out your applications of this method thoroughly in advance.

Behavior changes that you thought would meet a natural community of reinforcement, yet seem to still need generalization, perhaps have failed to awaken their natural community. See what response your student could use to awaken that community, estimate the rates and forms that will be effective but not counterproductive, and teach students how to use those forms at those rates, to see if the original behavior change still needs further support for generalization.

▶ Step 5: Teach Enough Examples

The most common mistake that teachers make, when they want to establish a generalized behavior change, is to teach one good example of it and expect the student to generalize from that example. After all, most of us are capable of generalizing from one good example.

Example: It would be good early training for teachers of generalizable concepts to have the assignment of producing generalized motor imitation in a few individuals with profound mental disabilities. Generalized motor imitation means that we can model any reasonable motor response for our student and expect a prompt and reasonably accurate imitation of it. Most persons with profound mental disabilities do not display this skill without special training. The training cannot be verbal explanation. Training consists of demonstrating a simple motor response and then hand-shaping a response in the student. The overall strategy that any teacher would propose would be to teach one example of this sort, and then another, and then another. Sooner or later, you expect a generalized effect. You expect that after some examples, you will demonstrate a new response to the student and have it imitated the first time, without it being hand-shaped over many laborious trials. Indeed, this will happen— in anywhere from 10 to 200 successive, hand-shaped examples of motor imitation, each one at least a little different from every other one. The first several of these examples may require hundreds, possibly even a few thousand, trials to accomplish. The last will require just one trial. Thus the total teaching time can be *very* long. The value of this example for teachers of generalizable concepts is the range of successive examples that can be necessary—from 10 to 200! Many teachers in the past expected to experience the small end of that range, but nothing like the large end of it. Consequently, many of them failed to produce gener- alized imitation after 20 or 30 examples and concluded that persons with profound mental disabilities are inherently incapable of such a complex, subtle skill. Later attempts by teachers prepared to try an almost indefi- nitely long series of examples succeeded in teaching well-generalized motor imitation (and then even generalized vocal imitation) quite con- sistently to those same persons who were previously thought incapable of any but the very simplest behavior changes. After they became capable of generalized motor imitation, they could *easily* be taught complex new responses (e.g., self-help skills) primarily by demonstration and a little supportive reinforcement.

> Behavior changes that do not generalize as expected should be taught again in another example, and then another, and then another. The teacher should always have prepared in advance a very long list of such examples, roughly rank-ordered for ease of acquisition, and they should always be prepared to keep teaching.

Now it is time for *you* to generalize.

The rule just stated says that it often will require more than one example of a lesson to produce a generalized behavior change in accordance with that les- son. This book has made use of teaching examples in an effort to produce generalized lessons in you, and it will continue to do so. The first time we made use of examples, we offered three of them. The second time, we used five. You might think that we followed our own advice. Ah, but the next time we exemplified an important point, we gave only one example (the girl whose wheelchair pushers had to be told to stop before she would generalize her newly acquired walking skill). And since then, we have offered only one example per point. Why? To provide you with a more valuable experience for furthering your generalization of these points than even a multiplicity of examples. That is the experience of *constructing your own further examples.*

- From now on, whenever you encounter a single example of any point, construct from your own experience, knowledge, or imagination a second, different example of that same point.

- First, go back to all the single examples of any point (they are labeled "Example") and construct, from your own experience, knowledge, or imagination a second, different example of that same point.

▶ **Step 6: Choose the Best Examples To Teach**

If you plan to establish a generalized behavior change by teaching enough examples of it to accomplish its generalization—and that is the oldest, most reliable, and most widely practiced generalization technique used—then you will need a very long list of teaching examples. Many such lists are arbitrary, except for trying to choose relatively easy ones to teach first and relatively difficult ones later, so as to maximize the student's experience of success. That is an excellent tactic. But so is task analysis. Think through the structure of the generalized behavior change that you want to produce to see what its components are. Be sure to represent those components in the list of examples, and try to maximize their representation as early as seems practical.

> **Example:** Persons with mental disabilities who were thought incapable of generalized imitation were taught it in almost the usual way. The exception was that their teaching examples were at first restricted deliberately to movements of the hand and fingers (e.g., hand-clapping, pointing, etc.). After a considerable number of examples of hand-and-finger imitations were taught, the students began to demonstrate generalized motor imitation. But the nature of their generalization was that they could now imitate almost any novel demonstration of a hand-and-finger response, but could not imitate any demonstrations of whole-body responses or vocal responses. Their behavior had generalized, but only within the realm of hand-and-finger imitations.
>
> Next, they were taught a series of whole-body imitations (getting up from their chairs, walking about the room, etc.), and after a considerable number of examples of this sort, they began to generalize more widely. Now they could imitate novel whole-body responses as well as novel hand-and-finger responses, but they still could not imitate vocal responses. They had to be taught a number of vocal imitations before generalized imitation of novel vocal imitations appeared. Other students were taught whole-body imitations first. When they generalized, it was only to novel whole-body responses, and not to hand-and-finger responses or vocal responses. Then, when hand-and-finger-imitations were taught, their generalization expanded to include imitation of novel hand-and-finger responses as well as novel whole-body responses. But they still did not generalize to vocal responses, which had to be taught next to produce generalization in that realm.

Example: Institutionalized children with mental disabilities were taught to dress themselves by teaching them successive examples of dressing skills. After they had been taught to put on underpants, they proved capable of putting on pants as well, without specific training in that skill, but no other clothes. After they had been trained to put on socks, they showed that now they could put on shoes as well, without specific training in that skill. But they could not yet put on the rest of their clothes. After they were then trained to put on shirts with sleeves and front buttons, they could also put on sleeved and sleeveless jackets, but could not put on sleeved or sleeveless pullovers. After training on sleeveless pullovers, they then could put on sleeved pullovers without specific training to do so, and thus were able to dress themselves (by institution standards).

The institutionalized children now could dress themselves. However, all of that training had been done by one ward aide. The children would dress themselves when he asked them to do so, but they would not do so for any other ward aide. A second ward aide, a woman, then began to repeat the training that the first ward aide had done. It went very quickly. Shortly after it was started, the children now began to dress for all the ward aides who dealt with them, and for new aides hired by the institution later on, men and women alike.

> When behavior changes are generalized by teaching successive examples of them, choose examples to represent the kinds of responses making up the generalized class, and the kinds of stimuli or situations in which that class is to be performed. Until those representatives are encountered in training, generalization of the training probably will not extend to what they represent.

Very often, what we teach is meant to be displayed for all other persons that our student ever meets. But almost as often, the student learns the behavior change that we have taught and uses it in our presence or with us, but with no one else. Because we want generalization across persons, then the successive examples that need to be taught are (at least) examples of persons. The teacher is the student's only example of a person, as far as this lesson is concerned. Thus we need more teachers, not to teach the lesson but to generalize it as was illustrated in the example about the institutionalized children who dressed themselves (at first) only for the ward aide who had taught them to dress themselves, but generalized to all ward aides when a second aide began to repeat that teaching.

Example: A small boy was considered unfriendly, even hostile, because he never showed social, affective response to other persons. One adult taught him to smile. The boy then smiled at that adult but at no one else. When it became obvious that smiling would not generalize, five more adults took turns teaching the boy to smile at each of them. This teaching went very quickly. Immediately after it had ended, the boy smiled at everyone he met and continued to do so. (Very likely, his smiling met a natural community of reinforcement.)

Thus, implicit in the general rule is to always teach successive examples that represent the dimensions of the generalization that we want. This is a smaller, but very useful, special case of that rule:

When behavior changes need to be generalized across persons, always be prepared to use more than one teacher. Add a second teacher, and if necessary a third, and if necessary more, until the lesson has not only been learned but also generalized to everyone else.

And take cheer from the experience of many studies of this technique. It is impressive how often just two teachers are enough to produce excellent generalization.

▶ **Step 7: Teach a Few Examples at the Same Time**

Usually, when we are teaching a behavior change that our student seems to find difficult, we are careful to teach one example of it at a time. Teaching more than one lesson at a time, we feel, intuitively, would surely be confusing to a student whose behavior is changing very slowly just with one lesson to learn. However, a very small but completely consistent set of studies suggests that whether or not learning the lessons goes more slowly, generalization from those lessons will occur more quickly if a few lessons are taught at the same time.

Example: Institutionalized children with mental disabilities who had almost no language behavior were chosen for language training. That training made no progress, however, because the children had very poor vocal imitation skills. A program aimed at improving their vocal imitation was substituted for the language training program. Sometimes, the vocal imitation training taught the children to imitate three new patterns of sounds, one pattern at a time. Each pattern was learned to a criterion of almost perfect imitation before the next pattern was taught. After all three had been learned, a test of how much the children had improved in their general ability to imitate new sounds was made. Sometimes, the vocal imitation training taught the children to imitate three new patterns of sounds, virtually at the same time. One or two teaching trials were conducted with one pattern, then one or two teaching trials were applied to the second pattern, then one or two were done with the third. This continued in a random way until all three had been learned to a criterion of almost perfect imitation. Then, a test of how much the children had improved in their general ability to imitate new sounds was made. Consistently, the children showed greater gains in generalized vocal imitation after they had just learned to imitate three new patterns at the same time, than after they had just learned to imitate three new patterns one at a time. (With their newly improved vocal imitation skills, the children were then returned to the language-training program and made good progress.) Incidentally, on the average, training the imitation of three new sound patterns at the same time took no longer than training the imitation of three new sound patterns, one at a time. The major source of difficulty in this training seemed to be the particular sounds involved in some patterns, rather than whether one or three were being learned at a time.

We have enough experience to suggest that it will be rewarding to try teaching more than one example at a time in many different behavior-change problems, but we do not yet have enough experience to know whether three at a time is better than two at a time or four at a time, and so on. It seems likely that a good number is a small number, but even that is a guess. You will have to explore this dimension of the technique yourself. Concern 1 (in the Some Concerns section) says that you should know how to do that. Please read Concern 1 again and take it seriously.

> When teaching the successive examples that represent the dimensions
> of the generalization that you want, try teaching those examples a few
> at a time, rather than one at a time. Measure whether the gain in gener-
> alization that this may well produce is worth any loss in learning
> progress that might result.

Teach Loosely?

We may sum up the meaning of these last several techniques in this way:

If you cannot teach every aspect of what you want learned, you have to teach examples of every aspect that you want learned. You probably have to represent all the dimensions of the final generalized behavior change that you want. And you might as well try to represent a number of them at the same time, rather than one for a time, and then another for a time. That might save time and is likely to further generalization. That will look like loose control of the teaching situation, but in fact, it might represent functional control of generalization.

If you take those principles to their logical outcomes, you arrive at the following pattern of procedural advice.

For whatever you are teaching:

- Use two or more teachers

- Teach in two or more places

- Teach from a variety of positions—sometimes in front of the student, sometimes beside the student, sometimes behind—sometimes sit, sometimes stand, sometimes kneel by the student

- Vary your tone of voice

- Vary your choice of words

- Show the stimuli from a variety of angles, using sometimes one hand and sometimes the other

- Have other persons present some times and not other times

- Dress quite differently on different days

- Vary the reinforcers

- Teach sometimes in bright light, sometimes in dim light

- Teach sometimes in noisy settings, sometimes in quiet ones

- In any setting, vary the decorations, vary the furniture, and vary their locations

- Vary the times of day when you and everyone else teach

- Vary the temperature in the teaching settings

- Vary the smells in the teaching settings

- Within the limits possible, vary the content of what is being taught

- Do all of this as often and as unpredictably as possible

We have no experience with teaching techniques as extreme as all that. We have a little experience that suggests that moving in that direction can enhance the generalization of the results of some teaching. A little experience with autistic children, and with a few other students who exhibit extreme stimulus over-selectivity, suggests that moving further in this direction might well overcome their extreme over-selectivity, such that they generalize better. Thus, the extreme pattern presented above is to illustrate what the extreme could look like, rather than to suggest that you operate right out there.

Remember the logic of teaching loosely. If you encounter generalization problems, consider loosening your teaching technique.

▶ Step 8: Make a Common Stimulus

It is very difficult to avoid stimulus control over a behavior change. (That is what this book is about.) That is just the way that behavior changes are. If there are any stimuli consistently present during the contingencies that shape a behavior change, then they are very likely to acquire control over that change. Present them, and the change is seen. In their absence, the change is not seen. The point of teaching loosely is to make many stimuli present during the shaping contingencies, for the following reasons:

1. No one stimulus, no small handful of stimuli, should acquire exclusive control over the behavior change and, if absent from other settings where the behavior change is desired, thereby preclude generalization to those settings.

2. There is a fair probability that any other setting where generalization is desired will contain at least some of the stimuli that were present during the shaping contingencies, such that these stimuli can encourage that generalization.

The second of these two arguments can be maximized by making sure that some important stimulus is common to both the teaching setting and any generalization setting. There are two tricks to that procedure. One is to be sure that the stimulus is important to the student; the other is to be sure that the stimulus is easily transportable to wherever generalization is wanted.

Example: A day-care center offered its children two major playtimes each day. After each playtime, the children were asked to put away the toys and materials with which they had been playing. The first playtime occurred when the teachers were available to help direct the children's putting away of the materials. But the second playtime occurred when the teachers were busy with another activity and needed to rely on the children to put away the things without teacher help. In fact, the children did not do much putting away, and the teachers found that they themselves did most of the first clean-up and all of the second clean-up— when they finally could get to the mess left after the second playtime. Analyzing the situation, they realized that during the first playtime, they were giving the children more attention for not putting things away than for helping. They reversed these contingencies, ignoring children who were not putting things away and praising children who were. This quickly produced very useful clean-up behavior in the children for each day's first playtime. This behavior change did not generalize to the second playtime. In the teachers' absence, the children continued their old ways of leaving the toys and materials on the floor when the second playtime ended. The teachers then brought to every first clean-up session a tape player loudly playing a tape on which, at random intervals of about 10 to 20 seconds, there was a distinctive brief tone. Of course, the children, hearing those tones, asked their teachers what they were, and the teachers replied, "Those tones tell us to put things away." Whenever a tone sounded, the teachers praised any child who was helping to clean up at that moment. After a little of this experience, the teachers made their praise more intermittent, praising children who were working at clean-up, but only after some tones, not after every tone, and never in the absence of a tone. As this was done, the children continued to clean up very well after the first playtime—and continued not to clean up after the second playtime. But then, one day, just as the second playtime ended, a teacher brought to the play area the tape player, turned it on, and left for her usual activities somewhere else. As the tape player began playing its tones, the children began cleaning up, even in the absence of a teacher. They continued to do so. Every day thereafter, the tape player was brought to the area and left there playing, and every day the children continued to clean up—without a teacher, but with an important functional stimulus common to their teaching setting and their generalization setting. The teachers had made that stimulus important (as a cue for the target behavior and as a cue for praise of the target behavior), and they had made it common.

> When a behavior change does not generalize from the setting where it was taught to other settings where it is desired, make a stimulus functional for it in the training setting that can be transported easily to the generalization setting.

Some stimuli make better common stimuli than others.

Example: A teacher was expanding the sight-word vocabulary of a primary grade boy who was far behind his peers in reading skills. The teacher conducted special sessions with him in a one-on-one setting away from the rest of the class, so as not to disturb them, bore them with material too easy for them, or emphasize the boy's delays relative to their attainments. The boy learned the new sight words quite easily in his private sessions with the teacher, yet never seemed to know them later when working with his classmates. The teacher then brought one of his peers from the class to help in his private sight-word sessions. The peer became a peer–tutor with the teacher guiding his tutoring. The boy then knew his sight-word vocabulary not only in the private sessions, but when he was with the rest of the class, as long as his peer–tutor was present (even though quietly present) in the class. The peer had been made a stimulus common to both settings and, by being a tutor in the private session, had been made an important functional stimulus there.

Tape-recorded tones are fine. They are cheap, easy to make, easy to make important, and easy to transport. Human beings are even better. They are free, already made, often already important, and, if not, they are easy to make important and present in many of the generalization settings that are important to us. In addition to all that, they are interesting, human, and members of those natural communities so important to really fundamental behavior changes and their elaboration and maintenance.

> When choosing a stimulus to be made common to both teaching and social generalization settings, consider human beings, especially in the role of peer–tutors.

▶ Step 9: Delay Reinforcement

Often, there is one stimulus very important to the teaching setting that is not present in the generalization settings—the reinforcer used for the teaching. The absence of the reinforcer in a generalization setting may be enough to preclude generalization to that setting. What better signal could there be that the target behavior will not be reinforced in the generalization setting than the absence of the reinforcer? One obvious and time-honored solution to this potential problem is to develop a schedule of reinforcement in the training setting that is so intermittent that the absence of the reinforcer in other settings will be almost indiscriminable. This solution, while often effective, nevertheless has a built-in disadvantage that sometimes defeats it. Very thin, very intermittent schedules of reinforcement, unless managed with great care, are likely to produce correspondingly low rates of the reinforced behavior. Thus, this technique may give you good generalization, but of so little behavior that you will not consider it a solution to the original problem.

Another way of making reinforcement less prominent a discriminative stimulus of the training setting, relative to generalization settings, is to delay it. But delayed reinforcement is not usually effective reinforcement, especially in the initial teaching of a new skill. However, in the maintenance of an already established skill, delayed reinforcement can be quite effective. And, it turns out, in the generalization of an already established skill, it can be quite useful.

> **Example:** During the two free-play periods of every preschool session, a young boy played well with the other boys, but almost never even talked to the girls. His teacher explained to him that he ought to play with both boys and girls, and explained why, and urged him to do so, but she had very little effect on his behavior pattern. Then, she told him that she would keep watch and count how many times he talked to or played with the girls. If he did that enough times, she would meet with him at the end of the preschool session, just before he went home, and let him choose a toy or sticker from a collection that she knew could be effective reinforcers for much of his behavior. If he did not play with the girls enough, she told him, she would say so when they met at the end of the session, and he would not get to choose anything from the collection.

In fact, the teacher could count the boy's interactions with all the other children only during the first free-play period of each session. His behavior during the second free-play period went uncounted by her (even though his behavior in both free-play periods was recorded by a research observer who did not communicate these results to the teacher). Thus she offered reinforcement, or withheld it, contingent on his behavior only during the first of the free-play periods of each session. But she met with him to give or withhold that reinforcement only after both free-play periods had occurred.

The result was that his play with girls generalized to the second free-play period, even though his behavior then made no difference as to whether or not he would receive his reinforcer later. Then, as an experiment, the teacher moved the time of meeting with the boy (to award or withhold reinforcement) to a time after the first free-play period (when his behavior, in fact, still determined the reinforcement), but before the second free-play period (which now obviously could not determine the reinforcement). Generalization of play with girls to the second free-play period promptly stopped. But resumption of the delayed timing after both free-play periods were over recovered it. (Eventually, play with girls met a natural community of reinforcement and no longer needed extrinsic reinforcement, immediate or delayed, to be maintained.)

> If an already established behavior change fails to generalize from its teaching setting to other settings, try delaying its reinforcement until a time later than the behavior will occur in those other settings. Explain what behavior the delayed reinforcement depends on, but not when it depends on that behavior.

▶ Step 10: Try Self-Monitoring and Verbal Mediation

If it can be effective to delay the possibility of reinforcement until after the desired behavior change has been displayed in its generalization settings, and then tell the student whether or not the student's behavior in those settings had earned the reinforcement—why not have the student tell you? That would mean that students had to become their own observers. They would have to observe their own behavior, remember or record it, and report on it accurately to their teacher. If it was adequate in the generalization settings, the teacher would award reinforcement, but otherwise would not. Rather than the teaching contingency operating in one setting but not another, it would, in one sense, operate in none of them. It would operate only on accurate self-reports by the students about their own generalization. In another sense, the contingency would operate everywhere—because it is observed everywhere—by the students themselves, who of course go everywhere that they go (which is something that their teacher, or any other observer, will usually find impractical to do). The crucial element in this technique is that the students report their own behavior accurately, or else discover that whenever they report it inaccurately, whether by accident or as a lie, their teacher nevertheless knows what the truth is. It may take some consistent teaching to establish that their teacher always knows the truth about their behavior in their generalization settings (even though the teacher apparently was not there). But once that is taught, truthful self-report seems to become a dependable behavior change itself. Thereafter, it can serve as a remarkably effective generalization technique.

Example: Preschool children were expected to share art materials during an unsupervised daily art activity period. Rather than sharing, they competed for the materials, hoarded them, and sometimes fought over them. A teacher explained to them that they must share and taught them how to share. But as soon as the teacher left, sharing stopped and the previous patterns of behavior returned. The teacher then told the children that he would meet with them after the art activity period and would ask them, one at a time, if they had shared and with whom. Those who said truthfully that they had shared enough would receive prizes of a sort that he knew were effective for many of their behaviors; the others would not. Some of the children reported inaccurately that they had shared enough; they were turned away with the gentle comment, "But you didn't really, did you?" Children who accurately reported sharing received reinforcement. Children who accurately reported not sharing were thanked for their report but not given anything else. (The teacher knew who had shared with whom, and who had not, because the children's behavior during the art activity period had been unobtrusively observed, recorded, and reported to the teacher just before his daily meeting with the children.) Very quickly, the children began reporting truthfully. Almost as quickly, they began sharing widely with each other so that they could later report truthfully that they had done so. This reporting was the only behavior that was reinforced in that way. Sharing itself was never reinforced, only true reports about it were. Thus the sharing that occurred was a kind of generalization from the direct teaching, which was teaching to tell the truth about sharing. Eventually, sharing seemed to meet a natural community of reinforcement and did not require reinforcement of its true reports.

Example: An overweight husband sought his wife's help in losing weight. His target behavior was to forgo snacks throughout the work day. (At home, he ate only reasonable meals that would not support his extra weight.) His wife gave him a golfer's counter to wear on his wrist like a wristwatch. His task, during his work day, was to count each urge to eat something that he resisted. His urges were private. He was the only person capable of counting them. When he came home at the end of his work day and could show his wife a large number on his wrist counter, she expressed pride, approval, and delight. When he displayed only small numbers, she simply said that she loved him, fat or slim, and talked pleasantly and normally about something else. His numbers steadily grew larger and larger, and his weight dropped. Eventually his numbers grew smaller and smaller, but his weight remained low. His explanation was that the small numbers meant that he now had very few urges to eat during the work day and so, when he resisted them, that made only small numbers. A friend asked why he lost weight, and he attributed it to his wife's delight in the large numbers that he could produce (at first) on his counter. His friend asked why he did not simply lie to his wife by counting falsely while still eating. He said that while she would have no way of knowing that any particular number was a lie, she certainly would be able to see over the long run whether he was losing weight, gaining weight, or holding steady. Thus, ultimately, she would know whether his reports generally were true or not. He stayed slim.

We have relatively little formal experience with these combinations of self-monitoring and verbal mediation to accomplish a generalized behavior change. What experience we have is impressive.

Try making students their own observers; reinforce their true reports about those observations when they show that the behavior change is occurring as desired. Do not use this technique unless you can be sure that you will reinforce only true reports.

The best way to conclude this book is to ask you to read again the subdivision titled Aim for a Natural Community of Reinforcement. It is the best of the techniques described here and, interestingly, it does not deserve the textbook definition of "generalization." It is a reinforcement technique, and the textbook definition of generalization refers to unreinforced behavior changes resulting from other, directly reinforced behavior changes. Also, please read again Concern 3 in the section entitled Some Concerns, to remind yourself that we are dealing with the pragmatic use of the word *generalization*, not the textbook meaning. We reinforce each other for using the word pragmatically, and it has served us well enough so far, so we will probably maintain this nonprecise usage.

Further Reading

Axelrod, S., & Hall, R. V. (1999). *Behavior Modification: Basic Principles.* Austin, TX: PRO-ED.

Catania, A. C. (1984). *Learning.* Englewood Cliffs, NJ: Prentice-Hall.

Hall, R. V. & Van Houten, R. (1983). *Behavior Modification: The Measurement of Behavior.* Austin, TX: PRO-ED.

Holman, J., and Baer, D. M. (1979). Facilitating generalization of on-task behavior through self-monitoring of academic tasks. *Journal of Autism and Developmental Disorders, 9*(4), 429–446.

Stokes, T. F., & Baer, D. M. (1977). An implicit technology of generalization. *Journal of Applied Behavior Analysis, 10,* 349–367.

Stokes, T. F., Fowler, S. A., & Baer, D. M. (1978). Training preshcool children to recruit natural communities of reinforcement. *Journal of Applied Behavior Analysis, 11,* 285–303.